Edward Augustus Inglefield

Birkenhead Iron-Clads

Edward Augustus Inglefield

Birkenhead Iron-Clads

ISBN/EAN: 9783337306632

Printed in Europe, USA, Canada, Australia, Japan

Cover: Foto ©berggeist007 / pixelio.de

More available books at **www.hansebooks.com**

BIRKENHEAD IRON-CLADS.

CORRESPONDENCE

BETWEEN

HER MAJESTY'S GOVERNMENT

AND

MESSRS. LAIRD BROTHERS;

AND

AN APPENDIX,

CONTAINING THE

CORRESPONDENCE BETWEEN OFFICERS OF H.M.'S
CUSTOMS AND CAPT. INGLEFIELD, R.N.,

AND

MESSRS. LAIRD BROTHERS,

RESPECTING

THE IRON-CLAD VESSELS

BUILDING AT BIRKENHEAD,

1863–4.

LONDON:

VACHER & SONS, 29, PARLIAMENT STREET.

Price One Shilling.

The following Letters form the entire Correspondence which has passed between Messrs. Laird Brothers and Her Majesty's Government, respecting the Iron-clad Vessels.

The production of these Letters, with other Papers, was moved for in the House of Commons by Mr. Seymour Fitzgerald, M.P., on the 23rd February last, but was refused by Her Majesty's Government.

They are now published with the permission of Messrs. Laird Brothers.

Vacher & Sons,
Publishers.

29, Parliament Street,
March, 1864.

CORRESPONDENCE

BETWEEN

HER MAJESTY'S GOVERNMENT

AND

MESSRS. LAIRD BROTHERS.

*Laird Brothers to S. Price Edwards, Esq., Collector of
H.M. Customs, Liverpool.*

[*Confidential.*]

BIRKENHEAD IRON WORKS, BIRKENHEAD,
4th September, 1863.

PRICE EDWARDS, ESQ.

Sir,—As the many rumours afloat in respect to
the two Iron Steam Rams built by us, and now lying
in our dock, have induced frequent and unusual visits
of Mr. Morgan, the Surveyor of Customs, to our
works, we are desirous of saving you any farther
unnecessary trouble about these vessels by giving you
our promise that they shall not leave the port without
your having a week's notice of our intention to deliver
them over to the owners, and we shall inform the
owners of this engagement on our part.

We may add that the first vessel will not be ready
for a month, and the second for six or seven weeks
from this date.

We are, Sir,
Your obedient Servants,
(Signed)   LAIRD BROTHERS.

A 2

*S. Price Edwards, Esq., Collector of H.M. Customs,
Liverpool, to Laird Brothers.*

CUSTOM HOUSE, LIVERPOOL,
5th September, 1863.

Gentlemen,—I beg to thank you for your note of
yesterday's date, wherein you kindly promise to
inform me, by a week's notice, of your intention to
deliver to the owners the two Iron Steam Rams now
being built by you, in order that the Government
may be informed of their being ready for sea.

This circumstance will, I know, be satisfactory to
the Board of Customs.

I am, Gentlemen,
Your obedient Servant,
(Signed)   S. PRICE EDWARDS.

Messrs. LAIRD BROTHERS.

———

H.M. Foreign Office to Laird Brothers.

FOREIGN OFFICE,
September 4th, 1863.

Gentlemen,—Earl Russell has been led to understand that you have intimated, that while you were
not in a position to volunteer information respecting
the Iron-clad Vessels, lately launched and now being
fitted out at your Yard, you would readily furnish information upon an official application, in writing,
being made to you for it.

Under these circumstances, Lord Russell has instructed me to request you to inform him, with as little delay as possible, on whose account, and with what destination these vessels have been built?

I am, Gentlemen,

Your obedient Servant,

(Signed)   A. H. LAYARD.

Messrs. LAIRD & Co.,
Birkenhead.

Laird Brothers to H.M. Foreign Office.

BIRKENHEAD,
September 5th, 1863.

A. H. LAYARD, ESQ., M.P.

Sir,—We have received your letter of the 4th instant, stating that Lord Russell has instructed you to request us to inform him, with as little delay as possible, on whose account, and with what destination we have built the Iron-clad Vessels recently launched, and now in course of completion at our Works.

In reply, we beg to say—that although it is not usual for Ship-builders to declare the names of parties for whom they are building vessels until the vessels are completed and the owners have taken possession— yet, in this particular case, in consequence of the many rumours afloat, coupled with the repeated visits of Mr. Morgan, the Surveyor of Customs, to our Works, we thought it right to ask permission of the parties on whose account we are building the vessels, to give their names to the English Government, in the event of such information being asked for officially, in writing.

They at once granted us the permission we sought for.

We therefore beg to inform you, that the firm on whose account we are building the vessels is— A. Bravay & Co., and that their address is No. 6 Rue de Londres, Paris, and that our engagement is to deliver the vessels to them in the Port of Liverpool when they are completed, according to our contract.

The time in which we expect to have the first vessel so completed, is not less than one month from this date, and the second, not less than six or seven weeks from this date.

We are, Sir,
Your obedient Servants,
(Signed)   LAIRD BROTHERS.

Laird Brothers to S. Price Edwards, Esq., Collector of H.M. Customs, Liverpool.
[*Confidential.*]

BIRKENHEAD IRON WORKS, BIRKENHEAD,
8th September, 1863.

S. PRICE EDWARDS, ESQ.

Sir,—Referring to our letter to you of the 4th instant, we think it right to inform you that it is our intention to take one of the Iron-clads—the " El Tousson "—from our graving dock for a trial trip on Monday next, within the usual limits of such trial trips; and you may rely on our bringing the vessel into the Birkenhead Float when the trial is finished, it being our intention to complete the vessel in the Birkenhead Float.

This trial is necessary, to test the machinery and other parts, but will not alter the time previously stated for the completion of the vessel.

We are, Sir,
Your obedient Servants,
(Signed)   LAIRD BROTHERS.

H.M. Treasury to Laird Brothers.

13,132$\frac{2}{9}$.

TREASURY CHAMBERS,
9th September, 1863.

Gentlemen,—I am desired by my Lords Commissioners of H.M. Treasury to acquaint you that their Lordships have felt it their duty to issue orders to the Commissioners of Customs, that the two Iron-clad Steamers now in course of completion in your Dock at Birkenhead, are not to be permitted to leave the Mersey until satisfactory evidence can be given of their destination, or at least until the inquiries which are now being prosecuted with a view to obtain such evidence shall have been brought to a conclusion.

I am, Gentlemen,
Your obedient Servant,
(Signed)   GEO. A. HAMILTON.

Messrs. LAIRD & Co.,
Birkenhead.

S. Price Edwards, Esq., Collector of H.M. Customs, Liverpool, to Laird Brothers.

CUSTOM HOUSE, LIVERPOOL,
11*th September,* 1863.

Dear Sirs,—I am sorry to say there can be no trial trip of the Iron-clad Ship until Earl Russell's reply can be had. That reply may yet come in time to meet your wishes.

I am, dear Sirs,
Yours very truly,
(Signed)   S. P. EDWARDS.

Messrs. LAIRD BROTHERS.

Laird Brothers to H.M. Treasury.

BIRKENHEAD IRON WORKS, BIRKENHEAD,
10*th September,* 1863.

GEORGE A. HAMILTON, ESQ.

Sir,—We are in receipt of your letter of the 9th instant, informing us that the Lords Commissioners of H. M. Treasury have issued orders to the Commisioners of Customs that the two Iron-clad Steamers now in course of construction by us, are not to be permitted to leave the Mersey until satisfactory evidence can be given of their destination.

In reply, we beg to inform you that we have

forwarded a copy of your letter to Messrs. A. Bravay & Co., at No. 6, Rue de Londres, Paris, on whose account we are building the vessels, and to whom we beg to refer you for further information, inasmuch as our engagement with them is to deliver the vessels at the Port of Liverpool when they are completed, according to our contract.

It may be useful to the Lords Commissioners of H.M. Treasury to know, that the time in which we expect to have the first vessel so completed is not less than one month from this date, and the second vessel not less than six or seven weeks from this date.

We are, Sir,

Your obedient Servants,

(Signed)   LAIRD BROTHERS.

S. Price Edwards, Esq., Collector H.M. Customs, Liverpool, to Laird Brothers.

LIVERPOOL,

14th *September*, 1863.

Dear Sirs, — You have the permission of the Government to try the Iron-clad Ship, on your guarantee to return her.

I have only this moment received the telegram.

Yours truly,

(Signed)   S. PRICE EDWARDS.

Messrs. LAIRD BROTHERS.

S. Price Edwards, Esq., Collector of H.M. Customs,
Liverpool, to Laird Brothers.

CUSTOM HOUSE, LIVERPOOL,
17th September, 1863.

Gentlemen,—With reference to the wish expressed by you that one of the Iron-clad Vessels in your Yard may be permitted to make a trial trip, I am directed to inform you that the Lords Commissioners of H.M. Treasury will allow the trial trip to be made by the vessel referred to in your letter of the 8th instant, relying upon the honourable engagement which has been given by you that the ship shall, after the usual trial trip, be brought back again to Liverpool, and shall not leave that port without a week's notice to H.M. Government of the intention to send her away.

I am, Gentlemen,
Your obedient Servant,
(Signed)   SAMUEL PRICE EDWARDS,
Collector.

Messrs. LAIRD BROTHERS,
Birkenhead.

Laird Brothers to S. Price Edwards, Esq., Collector of H.M. Customs, Liverpool.

BIRKENHEAD IRON WORKS, BIRKENHEAD,
18*th September*, 1863.

SAMUEL PRICE EDWARDS, ESQ.,
Collector H.M. Customs,
Liverpool.

Sir,—We beg to acknowledge receipt of your letter of the 17th inst. informing us that the Lords Commissioners of H.M. Treasury will allow the trial trip to be made by the vessel referred to in our letter of the 8th instant, relying upon the honourable engagement which has been given by us that the ship shall, after the usual trial trip, be brought back again to Liverpool, and shall not leave that port without a week's notice to H.M. Government of the intention to send her away.

This engagement was made under the circumstances set forth in our previous correspondence, and we now beg to confirm the same, and are,

Sir,
Your obedient Servants,
(Signed)   LAIRD BROTHERS.

H.M. Treasury to Laird Brothers.

TREASURY CHAMBERS,
19th September, 1863.

Gentlemen,—On the 13th instant the Lords Commissioners of H.M. Treasury directed the Board of Customs to inform you that their Lordships would allow a trial trip to be made by the Iron-clad Vessel referred to in a letter written by you on the 8th instant, in reliance upon the honourable engagement which had been given by your firm, that the vessel should, after the usual trial trip, be brought back again to Liverpool, and should not leave that Port without a week's notice to H.M. Government of the intention to send her away.

I am now commanded by the Lords Commissioners of H.M. Treasury to inform you that since that permission was given, circumstances have come to the knowledge of H.M. Government which give rise to apprehension that an attempt may be made to seize the vessel in question while on her trial trip.

I am to state to you explicitly that H.M. Government are convinced, that it is your intention, as far as it is in your power, to fulfil honourably the engagement into which you have entered; and that if any such attempt were made, it would be entirely without the privity of your firm, in whose good faith they place perfect confidence.

Inasmuch, however, as such an occurrence, in whatever method it may be brought about, would be contrary to the determination expressed by H.M. Government that the Iron-clad Vessels should be prevented leaving the Port of Liverpool until satisfactory evidence may be given as to their destination,

I am to state to you that this Board feel it their duty to apprize you that they cannot permit the trial trip except under provision against any forcible abduction of the vessels.

With this view, authority has been given to Admiral Dacres, who is now in the Mersey with the Channel Fleet, to place, with the concurrence of your firm, on board the Iron-clad Ram about to be tried, a sufficient force of seamen and marines in H.M. Naval Service to defeat any attempt to seize her. And I am to request that you will inform their Lordships whether you accept such assistance.

In the event of your refusing it, I am to inform you that the Board of Customs will be instructed to detain the vessel.

I am, Gentlemen,

Your obedient Servant,

The Messrs. LAIRD,   (Signed)   H. BRAND.

Birkenhead.

Laird Brothers to H.M. Treasury.

BIRKENHEAD IRON WORKS, BIRKENHEAD,
20th September, 1863.

H. BRAND, ESQ.

Sir,—We have the honour to inform you that we have received your letter of the 19th instant, and have been in communication with Admiral Dacres, and Mr. Edwards, Collector of Customs, on the subject, and will write to you to-morrow.

The trial of the Iron-clad Screw Steam Vessel is deferred.

We are, Sir,

Your obedient Servants,

(Signed)   LAIRD BROTHERS.

Laird Brothers to H.M. Treasury.

BIRKENHEAD IRON WORKS, BIRKENHEAD,
21st September, 1863.

THE HON. H. BRAND.

Sir,—We have the honour to reply to your letter of the 19th instant (received and acknowledged yesterday), informing us that circumstances have come to the knowledge of H.M. Government giving rise to an apprehension that an attempt may be made to seize our Iron-clad Steam Vessel on her trial trip, and stating that authority had been given to Admiral Dacres to place, with our concurrence, a sufficient force of seamen and marines on board her to defeat any such attempt.

We are not ourselves aware of any circumstance to induce us to entertain any such apprehension, but we beg to thank H.M. Government for the protection thus placed at our disposal, of which we shall gladly avail ourselves.

Owing, however, to what you have brought under our notice, and the incomplete state of the vessel, and also the present crowded state of the river Mersey, it will be desirable to defer the trial trip for some days; and, in the meantime, we trust that H.M. Government will be able to obtain further information as to any project that may exist to deprive us of our property.

We propose to communicate the substance of your letter to Messrs. A. Bravay & Co., of Paris.

We are, &c.,
(Signed)   LAIRD BROTHERS.

H.M. Treasury to Laird Brothers.

TREASURY CHAMBERS,
7th October, 1863.

Gentlemen,—Referring to your ready acceptance of the offer of H.M. Government to prevent any attempt at the forcible abduction of your property, the Iron-clad Vessel now nearly completed at Birkenhead, and understanding that the trial trip which has been the subject of former correspondence has been abandoned, I am directed by the Lords Commissioners of H.M. Treasury to acquaint you, that from information which has been received, it has become necessary to take additional means for preventing any such attempt.

Their Lordships have therefore given instructions that a custom-house officer should be placed on board that vessel, with full authority to seize her on behalf of the Crown, in the event of any attempt being made to remove her from the float or dock where she is at present, unless under further directions from their Lordships ; and likewise to obtain from the officer in command of H.M.S. " Majestic" any protection which may become necessary to support him in the execution of this duty.

My Lords request you to understand that these precautions are taken, not from any distrust of your intention to fulfil your engagement of giving a week's notice before the removal of the vessel, nor with the view of interfering in any way with your workmen in the completion of her, but exclusively for the purpose of preventing an attempt which may be made by other parties to nullify your engagement.

Their Lordships have directed Mr. Stewart, the Assistant-Collector of Customs at Liverpool, to communicate with you. And they doubt not that these precautions will meet with your concurrence.

I am, Gentlemen,
Your obedient Servant,
(Signed)   GEO. A. HAMILTON.

Messrs. LAIRD & CO.,
Birkenhead.

Laird Brothers to H.M. Treasury.

BIRKENHEAD IRON WORKS, BIRKENHEAD,
8th October, 1863.

G. A. HAMILTON, ESQ.

Sir,—We beg to acknowledge the receipt of your communication of the 7th instant, about the Iron-clad Vessel now nearest completion, and to inform you that we have been informed by Mr. W. G. Stewart, Assistant-Collector of H.M. Customs, Liverpool, that he has been directed to place a Customs officer on board the Iron-clad Vessel now nearest completion in the Great Float, Birkenhead, and that he has directions to seize her in case any attempt be made to remove her from where she is at present.

We have given the necessary order for admission to the vessel (called by us the "El Tousson") to Mr. Morgan, the Surveyor of Customs.

We are, respectfully,
Your most obedient Servants,
(Signed)   LAIRD BROTHERS.

Laird Brothers to H.M. Treasury.

BIRKENHEAD IRON WORKS, BIRKENHEAD,
9th October, 1863.

GEO. A. HAMILTON, ESQ.

Sir,—In further reply to your letter of the 7th instant (acknowledged yesterday), informing us that the Lords Commissioners of H.M. Treasury have given instructions that a Custom-house officer shall be placed on board the Iron-clad Vessel, now nearly completed, at Birkenhead, with full authority to seize her on behalf of the Crown in the event of any attempt being made to remove her from the float or dock where she is at present, unless under further directions from their Lordships, and likewise to obtain from the officer in command of H.M.S. "Majestic," any protection which may become necessary to support him in the execution of this duty. We beg to inform you that we have received this day a letter from Mr. Morgan, the Surveyor of Customs, giving us notice that, by direction of the Honourable Commissioners of Customs, he has this day seized the Iron-clad Vessel now lying in the Great Float at Birkenhead.

Since the receipt of your letter of the 7th instant no attempt has been made to remove the vessel from her moorings at the quay in the Great Float, and we are therefore at a loss to understand this apparent deviation from the decision of their Lordships, as expressed in their letter of the 7th, above referred to. But we consider this has been done, not with any distrust of our intentions to fulfil our engagement,

B

of giving a week's notice of our intention to remove the vessel, nor with the view of interfering in any way with the workmen in the completion of her, but exclusively for the purposes of preventing an attempt which may be made by other parties to nullify our engagement.

Although we are not aware of any circumstances to induce us to entertain any apprehension of any attempt being made to deprive us of our property by force, we gladly avail ourselves of any protection H.M. Government may think necessary for its security.

The vessel is still far from being ready for sea, and the work has been so much retarded by the excessively wet weather, that it will be some weeks before she is finally completed.

We are respectively,

Your obedient Servants,

(Signed)   LAIRD BROTHERS.

H.M. Treasury to Laird Brothers.

TREASURY CHAMBERS,
9th October, 1863.

$14,464\tfrac{9}{10}$.

Gentlemen,—I am commanded by the Lords Commissioners of H.M. Treasury, to inform you that, in consequence of information that has been received by H.M. Government as to the probability of a forcible abduction of one or both of the Iron-clad Vessels in course of completion in the Float at Birkenhead, their Lordships have felt it their duty to order the seizure of both these vessels, and have issued the necessary directions to the Commissioners of Customs accordingly.

I have the honour to be, Gentlemen,

Your obedient servant,

(Signed)   GEO. A. HAMILTON.

Messrs. LAIRD, Birkenhead.

Laird Brothers to H. M. Treasury.

BIRKENHEAD IRON WORKS, BIRKENHEAD,
17*th October*, 1863.

GEORGE A. HAMILTON, ESQ.

Sir,—In reply to your letter of the 9th instant, informing us that, "in consequence of information " which has been received by H.M. Government, as " to the probability of a forcible abduction of one or " both the Iron-clad Vessels in course of completion " in the Float at Birkenhead, their Lordships had felt " it their duty to order the seizure of both these " vessels, and have issued the necessary directions " to the Commissioners of Customs accordingly."

We have made the fullest inquiry, and have not been able to ascertain any circumstance to induce us to apprehend the probability of a forcible abduction of one or both of the Iron-clad Vessels in course of completion by us at Birkenhead—one, the " El Tousson," in the Great Float, the public dock, and the other, the "El Monnassir," in our own dock, on our own premises.

Both vessels are incomplete, and unfit for seagoing; the second vessel has not even got masts or funnel in, and both are in the sole charge of our own people.

We believe further, that if any such project as the forcible abduction of these vessels had ever been thought of, it could not successfully have been carried out in the Port of Liverpool.

Their Lordships have so often assured us that they are convinced that it is our intention, so far as in our power, to fulfil honourably the engagement which we have entered into with H.M. Government, that we have deferred making any formal protest against the seizure of these vessels, or the arbitrary and extraordinary measures that have been carried out in placing an armed force in charge.

We can only suppose that their Lordships have been induced to act as they have done by some information, which will be found, on further investigation, to have been entirely erroneous, or greatly exaggerated; and that they will, on the termination of the inquiries they have set on foot to investigate the case, feel justified in removing the vexatious restrictions they have placed upon our property, which have already caused and are still causing us an amount of loss and annoyance not easily estimated.

We remain, Sir, your obedient Servants,

(Signed)   LAIRD BROTHERS.

Laird Brothers to H.M. Treasury.

BIRKENHEAD IRON WORKS, BIRKENHEAD,
17*th October*, 1863.

GEORGE A. HAMILTON, ESQ.

Sir,—Referring to your letter of the 7th instant, in which you say that you understand the trial trip of the Iron-clad Steam Vessel the " El Tousson," now nearly completed, has been abandoned, we beg to inform you, that the trial trip was deferred but not abandoned, as you will see by referring to our letter addressed to the Hon. H. Brand, on the 21st September, in which we informed him that we considered it desirable to defer the trial trip of the first of the Iron-clads, the " El Tousson," in consequence of the then incomplete state of the vessel and the then crowded state of the River Mersey, and also in order that H.M. Government might be able to obtain further information as to the project which they had reason to apprehend was in existence—for seizing the Iron-clad Steam Vessel, by force, on her trial trip.

We now beg to inform you, that the work at the " El Tousson " is now in such a state of progress, as to make it desirable to have a trial trip to test the working of the machinery, and we, therefore, shall be glad to know, whether, with the information H.M. Government have been able to obtain since the date of our former letter, they still consider that the precautions of having a force of seamen and marines on board are necessary to protect our property.

We propose that the trial trip shall take place about the end of next week, or the beginning of the week after—say, some day between the 22nd and 29th instant—and that it should not extend beyond what is considered the limits of the Port or within sight of the Light-ship.

No circumstances have come to our knowledge to induce us to apprehend any attempt to take forcible possession of the vessel on her trial trip, and after the fullest inquiry we are satisfied, that if any such project ever existed in the Port of Liverpool, the real facts of the case would have been discovered before this, and the parties implicated placed under such surveillance as to render the execution of their design impossible.

Waiting your reply,

> We remain, Sir,
> > Your obedient Servants,
> > > (Signed)   LAIRD BROTHERS.

Laird Brothers to S. Price Edwards, Collector of H.M. Customs, Liverpool.

> BIRKENHEAD IRON WORKS, BIRKENHEAD,
> *19th October*, 1863.

S. PRICE EDWARDS, ESQ.

Sir,—Referring to the several communications we have had with you respecting the trial trip of the Iron-clad Steamer "El Tousson," now in course of completion in the Great Float, and the decision come to on or about the 21st September last, to postpone the trial trip until the work on board was in a more advanced state towards completion, we now beg to inform you, that the work is now in such a state of progress as to make it desirable to have a trial trip to test the working of the machinery, and we, therefore, shall be glad to know, whether, with the information H.M. Government have been able to obtain since the date of our former letter, they still consider that the precautions of having a force of seamen and marines on board are necessary to protect our property.

We propose that the trial trip shall take place about the end of this week or the beginning of the week after—say, some day between the 22nd and 29th instant—and that it should not extend beyond what is considered the limits of the Port or within sight of the Light-ship.

No circumstances have come to our knowledge to induce us to apprehend any attempt to take forcible possession of the vessel on her trial trip, and after the fullest inquiry, we are satisfied that if any such project ever existed in the Port of Liverpool, the real facts of the case would have been discovered before this, and the parties implicated placed under such surveillance as to render the execution of their design impossible.

We remain, Sir,
Your most obedient Servants,
(Signed)   LAIRD BROTHERS.

H.M. Treasury to Laird Brothers.

[*Immediate.*]

TREASURY CHAMBERS,
21*th October*, 1863.

15,023$\frac{21}{10}$.

Gentlemen,—In reply to your letter of the 17th instant, relating to the Iron-clad Vessels which you are fitting out, I am commanded by the Lords Commissioners of H.M. Treasury that, after duly weighing all the circumstances of the case, H.M. Government are unable to consent to the trial trip of one of those vessels, the " El Tousson," taking place, as proposed by you; neither can they allow the removal of the armed force which is stationed for the purpose of upholding the Custom House officers in possession of the vessel.

I am, Gentlemen, your obedient Servant,

(Signed)   GEO. A. HAMILTON.

Messrs. LAIRD,
    Birkenhead Iron Works, Birkenhead.

Laird Brothers to H.M. Treasury.

BIRKENHEAD IRON WORKS, BIRKENHEAD,
24*th October*, 1863.

GEORGE A. HAMILTON, ESQ.

Sir,—We beg to acknowledge receipt of your letter of 21st instant, in which you inform us that H.M. Government, after duly weighing all the circumstances of the case, are unable to consent to the trial trip of one of the vessels, the " El Tousson," taking place, as proposed by us.

We beg to state that we did not propose that the trial trip should take place under any other conditions than were set forth in their Lordships' letter of the 19th September, unless from information received since the date of that letter their Lordships should think it no longer necessary to place a force of seamen and marines on board to protect our property; on the contrary, if H.M. Government still apprehend any attempt, we will gladly avail ourselves, as already stated in our letter of 21st September, of any protection H.M. Government may think necessary to defeat any such attempt.

We therefore respectfully renew our application to make the trial trip in the course of next week, or within any other suitable time.

We are, Sir, your obedient Servants,
(Signed)   LAIRD BROTHERS.

H.M. Treasury to Laird Brothers.

[*Immediate.*]

TREASURY CHAMBERS,
27th October, 1863.

15,310$\frac{27}{10}$.

Gentlemen,—In reply to your letter of 24th inst., I am commanded by the Lords Commissioners of H.M. Treasury to acquaint you, that they are unable to comply with your request to make a trial trip of the "El Tousson," one of the Iron-clad Vessels fitting in your yard at Birkenhead, in the course of this week, or within any other suitable time.

I am, Gentlemen, your obedient Servant,
(Signed)   GEO. A. HAMILTON.

Messrs. LAIRD BROTHERS,
    Birkenhead Iron Works, Birkenhead.

S. Price Edwards, Esq., Collector of Customs, Liverpool,
to Laird Brothers.

CUSTOM HOUSE, LIVERPOOL,
27th October, 1863.

Gentlemen,—I hereby beg to inform you, that your two Cupola Vessels are now detained, under the 223 section of "The Customs Consolidated Act," the ground of detention being a violation of "The Foreign Enlistment Act." And I take leave further to state, that the officers in charge have received directions to remove your workmen at once from on board the ships.

I am, Gentlemen, your obedient Servant,
(Signed)　S. PRICE EDWARDS,
Collector.

Messrs. LAIRD BROTHERS,
Birkenhead.

———

Laird Brothers to H.M. Foreign Office.

(TELEGRAM.)　*29th October,* 1863.

From LAIRD BROTHERS, Birkenhead, *to* EARL RUSSELL,
Foreign Office, Whitehall, London.

Captain Inglefield informs us that his orders are to take the two Iron-clads into the river Mersey. We protest against the probable destruction of our property in having ships (one of which is a mere hulk without masts, funnel or stearing gear) taken out of docks, where they are now in safety, and moored in the river at this inclement season of the year, and we trust that the orders sent to Captain Inglefield will be reconsidered.

Same sent to
G. A. HAMILTON, Secretary to the Treasury,
Treasury, Whitehall, London;
And to the Secretary to the Admiralty,
Whitehall, London.

H.M. Treasury to Laird Brothers.

(TELEGRAM.) *29th October*, 1863.

From the ASSISTANT-SECRETARY, the Treasury, White-hall, *to* Messrs. LAIRD, Birkenhead.

Captain Inglefield will no doubt, in his dispositions regarding the Iron-clad Vessels, take every proper precaution for the preservation of the property. The orders have been well considered and cannot be revoked or altered.

Laird Brothers to H.M. Foreign Office, Treasury and Admiralty.

BIRKENHEAD IRON WORKS, BIRKENHEAD,
29th October, 1863.

THE RIGHT HON. EARL RUSSELL.

My Lord,—We sent you this morning the following telegram:—

" Captain Inglefield informs us that his orders " are to take the two Iron-clads into the river " Mersey.

" We protest against the probable destruction of " our property in having ships (one of which is a " mere hulk without masts, funnel or stearing gear) " taken out of docks, where they are now in safety, " and moored in the river at this inclement season " of the year, and we trust that the orders sent to " Captain Inglefield will be reconsidered," which we now beg to confirm.

We are, my Lord, your most obedient Servants,
(Signed) LAIRD BROS.

Same sent to
   G. A. HAMILTON, Esq., Secretary to the Treasury;
and to
   The Secretary to the Admiralty.

Laird Brothers to H.M. Treasury.

BIRKENHEAD IRON WORKS, BIRKENHEAD,
29th October, 1863.

THE LORDS COMMISSIONERS OF
H.M. TREASURY.

My Lords,—We beg to call your Lordships' attention to the very serious position in which we are placed by the extraordinary steps taken by H.M. Government with the two Iron-clad Ships now being built by us.

It is a rule well recognised in all trading establishments, that an order, whilst under execution, is the property of the person giving it, and that a builder has no right to make public the orders or instructions of his employers. This is a rule of business which must be well known to H.M. Government.

On the 4th September, however, we were officially applied to by Mr. Layard for the information—"on whose account, and with what destination, the vessels were being built."

Owing, however, to certain vague rumours which were current in the newspapers, and to the repeated visits of Mr. Morgan, the Surveyor of H.M. Customs, we had taken the precaution to obtain the owner's sanction to disclose his name, and we were accordingly enabled, by return of post, to reply to Mr. Layard's letter, and inform him that we were building the ships for Messrs. A. Bravay & Co., 6, Rue de Londres, Paris.

On the 9th September, Mr. Hamilton, the Secretary to the Treasury, wrote us to say that the vessels would not be permitted to leave the Mersey until inquiries then being prosecuted had been brought to a conclusion.

In order to give H.M. Government ample time to make these inquiries, we wrote in reply to say that the first vessel would not be complete in less than a month. And about the same time we stated that the first vessel would be ready for a trial trip in a short time, and that we would engage that she should return to the Birkenhead Float.

On the 17th September, permission was given for the trial trip, and we were further requested to give our personal undertaking that the vessel should not leave the port without our giving a week's notice to H.M. Government.

This undertaking we readily gave by return of post.

On the 19th September, we received a letter from Mr. Brand, Secretary of the Treasury, to say that the Government feared an attempt might be made to seize the vessel whilst on her trial trip, but without giving any reason for such apprehension, and tendering the services of a force of seamen and marines.

We accepted this offer of protection, though unable, ourselves, to discover any grounds for such apprehension.

On the 7th October, we received a letter from Mr. Hamilton, Secretary of the Treasury, stating that, from further information, it had become necessary that a Custom House Officer should be placed on board, and the Captain of the " Majestic " was instructed to afford him protection.

As none of these movements of H.M. Government

interfered with us in our completion of these ships, and as any plans to seize our ships, either by the Northern or Southern belligerents, would entail great pecuniary loss upon us, we, of course, made no objection to these means provided by the Government for our protection, though we were then, and still are, unable to discover any grounds whatever for these precautionary measures, and we are satisfied that H.M. Government have leant too credulous an ear to the inventions of designing persons.

But when H.M. Government—without giving us any information to show us that they have any just grounds for doing so — proceed to seize our ships and turn off our workmen, and threaten to remove a helpless hulk from a place of safety into the open roadstead of the Mersey, we feel it our duty to enter our indignant protest against proceedings so illegal and so unconstitutional.

We have dealt candidly and openly with H.M. Government. We have, with the owners' permission, given the names of the owners, and we believe we have a perfect legal right to build ships for a French subject without requiring from him a disclosure of his object in having such vessels constructed. It forms no part of our duty to interfere in any way with his affairs, and we shall not do so.

We need hardly say that we hold the Government responsible to us for the large pecuniary loss we shall sustain by these arbitrary proceedings.

We are, My Lords,
Your Lorships' most obedient Servants,
(Signed)   LAIRD BROTHERS.

Laird Brothers to H.M. Treasury.

BIRKENHEAD IRON WORKS, BIRKENHEAD,
30th October, 1863.

THE SECRETARY TO THE TREASURY.

Sir,—We received your telegram late yesterday afternoon, as follows:—

" Captain Inglefield will no doubt, in his disposi-
" tions regarding the Iron-clad Vessels, take every
" proper precaution for the preservation of the pro-
" perty.   The orders have been well considered, and
" cannot be revoked or altered."

We take the liberty to draw the attention of H.M. Government to the peculiar construction of the hulls and machinery of the vessels built by us and seized by the Government, and to express to them our conviction that it is not possible in their present incomplete state for any Naval officer by any dispositions he can make to protect the vessels from damage even in a dock, and much less in the open roadstead of the Mersey, where, in our opinion, they cannot even be moored with safety in this inclement season of the year.

We are, Sir,
Your obedient Servants,
(Signed)   LAIRD BROTHERS.

H.M. Foreign Office to Laird Brothers.

FOREIGN OFFICE,
October 30th, 1863.

Gentlemen,—I am directed by Earl Russell to acknowledge the receipt of your letter of the 29th instant, containing copy of a Telegraphic Message which you addressed to his Lordship on that day, protesting against the removal into the Mersey of the two Iron-clad Vessels now under seizure by the Crown, and I am to state to you that the matter has been referred to the Lords Commissioners of H.M. Treasury.

I am, Gentlemen,
Your most obedient Servant,
(Signed)   E. HAMMOND.

Messrs. LAIRD,
Birkenhead.

H.M. Admiralty to Laird Brothers.

<table>
<tr><td>In replying, quote the following
Initial Letter
M.</td><td>ADMIRALTY,
30th October, 1863.</td></tr>
</table>

Gentlemen,—I am commanded by my Lords Commissioners of the Admiralty to acknowledge the receipt of your letter of the 29th instant, forwarding a copy of the Telegram sent by you to this Office yesterday morning, and confirming its contents, by which you protest against the two Iron-clads being taken into the River Mersey, and request that the orders given to Captain Inglefield may be reconsidered.

In reply I am to inform you that my Lords have referred your Telegram to the Secretary of State for Foreign Affairs.

I am, Gentlemen,
Your most obedient Servant,
(Signed)   W. G. ROMAINE.

Messrs. LAIRD BROTHERS,
Birkenhead.

Laird Brothers to H.M. Foreign Office and Treasury.

(TELEGRAM.) 31*st October*, 1863.

From LAIRD BROTHERS, Birkenhead, *to* EARL RUSSELL, Foreign Office, Downing Street, London.

Admiralty write, that they have referred to the Secretary of State for Foreign Affairs our telegraph and letter of twenty-ninth instant, protesting against the removal of the Iron-clads into the river Mersey. We renew our protest against moving the vessels from the Dock, where they are now in perfect security either from forcible abduction or sea risk. The weather is now most boisterous, and always uncertain at this time of the year.

A similar Telegram sent to
   The Secretary to the Treasury.

H.M. Foreign Office to Laird Brothers.

FOREIGN OFFICE,
2nd November, 1863.

Gentlemen,—I am directed by Earl Russell to acknowledge the receipt of your telegraphic message of the 31st ult., renewing your protest against the removal from the Docks of the two Iron-clad Vessels now under seizure by the Crown; and I am to state to you that the matter has been referred to the Lords Commissioners of H.M. Treasury.

I am, Gentlemen,
    Your obedient Servant,
      (Signed) E. HAMMOND.

Messrs. LAIRD,
   Birkenhead.

C

H.M. Treasury to Laird Brothers.
Immediate.

$15,586\frac{2}{11}$.

TREASURY CHAMBERS,
2nd November, 1863.

Gentlemen,—In reply to your letter of the 30th ult., representing the danger which the Iron-clad Vessels now under seizure will incur in consequence of their being removed from your premises, I am commanded by the Lords Commissioners of H.M. Treasury to acquaint you, that their Lordships having been in communication with the Secretary of State in reference to your letter, have nothing to add to the communication made to you by telegram on the 29th ult.

I am, Gentlemen,
Your obedient Servant,
(Signed) GEO. A. HAMILTON.

Messrs. LAIRD,
Birkenhead.

H.M. Treasury to Laird Brothers.
Immediate.

$\left.\begin{matrix} 15,467 \\ 15,608 \\ 15,646 \end{matrix}\right\}\frac{3}{11}$.

TREASURY CHAMBERS,
3rd November, 1863.

Gentlemen, — I am commanded by the Lords Commissioners of H.M. Treasury, to acknowledge the receipt of your letter of 29th ult., in which you protest against the measures which H.M. Government have felt it their duty to adopt for effecting and maintaining the seizure of the two Iron-clad Vessels at Birkenhead.

I am desired to state, in reply thereto, that their Lordships can only refer you to the notice of the cause of seizure conveyed in the letter addressed to your firm by the Collector of Customs, on the 27th ult., and that their Lordships must decline to enter into any discussion of the subject with you before the investigation which the case will necessarily receive in a court of law.

I am, Gentlemen,

Your obedient Servant,

(Signed) GEO. A HAMILTON.

Messrs. LAIRD BROTHERS,
Birkenhead.

Laird Brothers to H.M. Foreign Office, Treasury and Admiralty.

BIRKENHEAD IRON WORKS, BIRKENHEAD,
7th December, 1863.

THE RIGHT HON. EARL RUSSELL.

My Lord,—We beg to call your attention to the present condition of the two Steam Vessels, the " El Tousson" and the " El Monnassir," which have been removed by Captain Inglefield 'from dock into the river Mersey.

On Thursday last it blew a very heavy gale of wind here, and several large vessels, one of them a large steamer, were driven from their moorings within the estuary.

We understand that no steps are as yet taken to bring the rights of the Crown before a Jury, and in the meantime the vessels are exposed to great risk.

It is a matter of serious importance to us, as in case the vessels should be lost or burned in the Mersey before we can deliver them to the owners, we shall be thereby prevented from completing our contract.

Our attention is more immediately called to this subject by the fact that one of the fire policies on the " El Monnassir" expires to-day, and we are in doubt what, under the circumstances, we ought to do.

It is evident that the vessels ought to be insured, both against sea-risk and fire, and we shall be glad to know whether H.M. Government have taken these precautions for the security of the property, and if not, whether they intend to do so?

We may further state, that we trust the Government have given strict orders that proper precautions are taken for the preservation of the property from the injury and deterioration it is liable to from exposure to the damp and wet at this inclement season.

We are, my Lord,
Your obedient Servants,
(Signed)   Laird Brothers.

Copy of above to
    Secretary to Treasury, same date;
    Secretary to Admiralty, ditto.

H.M. Treasury to Laird Brothers.

17,571$\frac{10}{12}$.

TREASURY CHAMBERS,
10th December, 1863.

Gentlemen,—I am desired by the Lords Commissioners of H.M. Treasury to acknowledge the receipt of your letter, dated 7th instant, calling their

Lordships' attention to the exposed condition of the two Steam Vessels, "El Tousson" and "El Monnassir," now lying in the river Mersey, under charge of Captain Inglefield, R.N.

Your letter, though dated the 7th instant, was not received in London till the 10th, with the Birkenhead postmark of the 9th instant.

My Lords desire me to acquaint you that the subjects adverted to in your letter shall receive immediate attention, and that a further communication will be addressed to you thereon.

I am, Gentlemen,
Your obedient Servant,
(Signed) GEO. A. HAMILTON.

Messrs. LAIRD BROTHERS,
  Birkenhead.

H.M. Foreign Office to Laird Brothers.

FOREIGN OFFICE,
December 11th, 1863.

Gentlemen,—I am directed by Earl Russell to acknowledge the receipt of your letter of the 7th instant, which was only received at this Office on the 10th; and I am to inform you that your letter has been referred to the Treasury, for such directions as that Department may think proper to give.

I am, Gentlemen,
Your most obedient Servant,
(Signed) E. HAMMOND.

Messrs. LAIRD BROTHERS,
  Iron Works, Birkenhead.

The Admiralty to Laird Brothers.

In replying, quote the following Initial Letter
M.

ADMIRALTY,
12th December, 1863.

Gentlemen,—I have received and laid before My Lords Commissioners of the Admiralty your letter of the 7th instant, calling attention to the present condition of the two Steam Vessels, "El Tousson" and "El Monnassir," and as to the preservation of the property on board from injury, &c.

I am, Gentlemen,
Your most obedient Servant,
(Signed)   C. PAGET.

Messrs. LAIRD & Co.,
Birkenhead.

H.M. Treasury to Laird Brothers.

$\left.\begin{array}{l}18,046 \\ 17,571\end{array}\right\}\frac{18}{12}$

TREASURY CHAMBERS,
18th December, 1863.

Gentlemen,—With further reference to your letter of the 7th instant, respecting the present condition of the two Steam Vessels, "El Tousson" and "El Monnassir," I am desired by the Lords Commissioners of H.M. Treasury to acquaint you, that it is the intention of H.M. Government that the existing insurances on these vessels should be kept up or renewed, *ad interim*, at the cost of the public, and in the name of some person on Her Majesty's behalf, who, if you will agree to repay the cost of such insurance, in the event of the property in the vessels being hereafter adjudged to you, may be constituted a trustee of the policies for H.M., or for such person or persons as may hereafter be adjudged to be the owner or owners of the vessels, according to the result of the proceedings which may be taken for the purpose of deciding on the validity of the seizures.

As regards the precautions to be taken for preserving the vessels from injury by weather, My Lords are satisfied that every possible precaution has been already taken, and will continue to be taken, by the Naval officer in command at Liverpool, and that no deterioration of any kind need be anticipated.

I am, Gentlemen,

Your obedient Servant,

(Signed)   GEO. A. HAMILTON.

Messrs. LAIRD BROTHERS,
    Birkenhead.

Laird Brothers to H.M. Treasury.

BIRKENHEAD IRON WORKS, BIRKENHEAD,
21st December, 1863.

G. A. HAMILTON, ESQ.,

Sir,—We beg to acknowledge the receipt of your letter of the 18th instant, and hope to send a reply to-morrow.

We are, Sir,

Your obedient Servants,

(Signed)   LAIRD BROTHERS.

Laird Brothers to H.M. Treasury.

BIRKENHEAD IRON WORKS, BIRKENHEAD,
22nd December, 1863.

THE SECRETARY TO THE TREASURY.

Sir,—We have the honour to acknowledge the receipt of your letter of the 18th instant, stating that it is the intention of H.M. Government to keep up and renew, *ad interim,* the insurances of the " El Tousson" and " El Monnassir," at the cost of the public, provided we will agree to repay the cost of such insurances, in the event of the property in the vessels being hereafter adjudged to us, according to the result of the proceedings which may be taken for the purpose of deciding on the validity of the seizures.

In reply we beg respectfully to submit to you that the condition we are asked to agree to is not reasonable.

For, not only do the vessels incur marine risk by being exposed in the Estuary of the Mersey, which risk would not have arisen if the vessels had remained in the Docks, but the time has expired during which they would have been in our possession at all.

If they had remained in dock no marine insurance would have been necessary, and if they had not been seized, they would ere this have been delivered to the purchasers.

Under these circumstances we respectfully submit that the vessels should be insured, and kept insured, at the public cost, without any such condition being imposed on us.

We beg to inform you that another Policy against fire for £20,500 expires on the 24th instant.

We are, Sir, your obedient Servants,

(Signed) LAIRD BROTHERS.

P.S.—Since writing the above we find that two further Policies against Fire, one for £14,000 and another for £5,000, also expire on the 24th instant.

(Signed)   LAIRD BROTHERS.

Laird Brothers to H.M. Treasury.

BIRKENHEAD IRON WORKS, BIRKENHEAD,
30*th December*, 1863.

THE SECRETARY TO THE TREASURY.

SIR,—We beg respectfully to draw your attention to our letter of the 22nd instant respecting the insurance on the "El Tousson" and "El Monnassir," and to request an early reply.

We are, Sir,

Your obedient Servants,

(Signed)   LAIRD BROTHERS.

Laird Brothers to H.M. Treasury.

BIRKENHEAD IRON WORKS, BIRKENHEAD,
9*th January*, 1864.

THE SECRETARY TO THE TREASURY.

Sir,—We wrote to you on the 30th December, drawing attention to our letter of 22nd December, respecting the insurance on the "El Tousson" and "El Monnassir," and asking a reply thereto.

As we have not yet received any communication on the subject, we would again respectfully ask an early reply.

We are, Sir,

Your obedient Servants,

(Signed)   LAIRD BROTHERS.

Laird Brothers to H.M. Treasury.

BIRKENHEAD IRON WORKS, BIRKENHEAD,
12th January, 1864.

THE SECRETARY TO THE TREASURY,

Sir,— From communications which have passed between our solicitors and those of the Government, in the case of the "El Tousson" and " El Monnassir," it would appear that the trial may not come on for a considerable time, and consequently the vessels will have to remain in their present exposed position, unless some other arrangement can be made with the Government.

Were the vessels finished, there would be much less risk of their receiving damage than in their present unfinished and unprotected state.

We therefore think it desirable to make the following proposals to the Government, namely — that the vessels should be moved into the Birkenhead public docks, and placed at the top end of the great float, about a mile from the entrance, the Government retaining possession by an armed force, or otherwise, as they may think requisite, so that we may be able to complete our contract, which we are desirous of doing, although the value of the additional fittings we should put on board would be very considerable.

In the event of the Government proving their right to retain the vessels, they will, if our proposal be agreed to, be in a much more perfect state. On the other hand, should the Government not succeed, the vessels will be sooner ready for delivery by us to the owners, and consequently any claim for damages against the Government would be reduced.

These proposals are made without prejudice to any legal proceedings Messrs. A. Bravay & Co., or ourselves may be advised to take, for obtaining compensation in this matter, and being advantageous to H.M. Government we hope they will accede to them.

We desire further, to add, that we have no hesitation in saying these vessels will be much more secure in the great Float than they now are in the river Mersey; and in support of this opinion we enclose a plan showing where the vessels are at present moored, and where we purpose to have them placed.

We are, Sir,

Your most obedient Servants,

(Signed)   LAIRD BROTHERS.

The plan enclosed indicates the various positions of the vessels " El Tousson " and " El Monnassir," as follows :—

1. Situation in the Birkenhead Dock, where "El Tousson" was lying when seized.

2. Messrs. Laird Brothers' Dock, where the " El Monnassir " was lying when seized.

3. Present position of " El Tousson " and " El Monnassir " in river Mersey.

4. Situation in Birkenhead public Dock where it is proposed, by Laird Brothers' letter of 12th January, to place the vessels for completion.

H.M. Treasury to Laird Brothers.

570 $\frac{13}{1}$

TREASURY CHAMBERS,
14*th January*, 1864.

Gentlemen,—In reply to your letter of the 12th instant, proposing that the " El Tousson " and " El Monnassir " should be placed in the Birkenhead Docks, and there completed, I am commanded by the Lords Commissioners of H.M. Treasury to inform you that their Lordships regret that they are unable to comply with your request.

I am, Gentlemen,
Your obedient Servant,
(Signed)   GEO. A. HAMILTON.

Messrs. LAIRD & CO.,
   Birkenhead.

H.M. Treasury to Laird Brothers.

209 } $\frac{20}{1}$
464 }

TREASURY CHAMBERS,
20*th January*, 1864.

Gentlemen,—In reply to your letter of the 9th instant and previous letters, I am commanded by the Lords Commissioners of H.M. Treasury to acquaint you, that their Lordships will provide in the manner they may consider requisite, against the risks from fire and other damages to the Iron-clad Vessels " El Tousson " and " El Monnassir " while they remain in possession of H.M. Government.

I am, Gentlemen,
Your obedient Servant,
(Signed)   GEO. A. HAMILTON.

Messrs. LAIRD,
   Birkenhead Iron Works, Birkenhead.

Laird Brothers to H.M. Treasury.

BIRKENHEAD IRON WORKS, BIRKENHEAD,
25th January, 1864.

THE SECRETARY TO THE TREASURY.

Sir,—We have the honour to acknowledge the receipt of your letter of the 14th instant, refusing us permission to finish the ships " El Tousson " and " El Monnassir "; and also of your letter of the 20th instant, stating that the Lords of the Treasury will provide in the manner they may consider requisite against the risks of fire and other damage.

If this decision about completing the ships be adhered to, we shall be prevented for an indefinite period from completing our contract, and, consequently, be kept out of a very large sum of money, which will be due to us, and which the owners are ready to pay to us, as soon as the vessels are so completed and delivered to them in the Port of Liverpool.

As stated in our former letter—we are perfectly satisfied that the Government should retain possession by an armed force, or otherwise, as they may think requisite, until the legal proceedings now pending are terminated, or some other settlement of the question arrived at.

Taking all these circumstances into consideration, we trust that their Lordships may see reason to alter their decision and agree to the proposal contained in our letter of the 12th instant.

In the meantime, we beg to call the attention of the Lords of the Treasury to the fact that, though it is now several months since the vessels were seized, yet no steps have as yet been taken to bring the matter to a legal decision, although our attorneys have repeatedly pressed this course on the Law Advisers of the Crown.

We are, Sir,

Your obedient Servants,

(Signed)   LAIRD BROTHERS.

H.M. Treasury to Laird Brothers.

TREASURY CHAMBERS,
2nd February, 1864.

Gentlemen,—In reply to your further letter of 25th ultimo, I am desired by the Lords Commissioners of H.M. Treasury to acquaint you that H.M. Government cannot permit the Iron-clad Vessels built in your Yard and now under seizure, to be completed.

I am, Gentlemen,

Your obedient Servant,

(Signed)   GEO. A. HAMILTON.

Messrs. LAIRD BROTHERS,
Birkenhead Iron Works, Birkenhead.

Laird Brothers to H.M. Treasury.

BIRKENHEAD IRON WORKS, BIRKENHEAD,
3rd *February*, 1864.

THE SECRETARY TO THE TREASURY.

Sir,—We are in receipt of your letter of the 2nd instant, in which you inform us that H.M. Government cannot permit the Iron-clad Vessels built in our Yard, and now under seizure, to be completed.

We beg however to call your attention to the fact that no information has yet been afforded to us in reply to our repeated requests to know when the legal proceedings in the Court of Exchequer will be brought to trial before a Jury.

We are informed by our legal advisers that they have repeatedly pressed this matter on the attention of the Law Officers of the Crown, but are unable to obtain any satisfactory information, although the case might have been brought to trial in November last, or in January last.

We therefore feel ourselves entitled to urge upon H.M. Government the propriety of their at once informing us as to the time when they purpose to bring this matter to trial.

It must be apparent that this continued delay in bringing the matter to a legal issue is an act of injustice to ourselves and to the owners of the ships.

We are, Sir,
Your obedient Servants,
(Signed)   LAIRD BROTHERS.

H.M. Treasury to Laird Brothers.

Immediate.

2,185$\frac{8}{2}$

TREASURY CHAMBERS,
8th February, 1864.

Gentlemen,—In reply to your letter of 3rd instant, I am commanded by the Lords Commissioners of H.M. Treasury to acquaint you that they are informed that an "information" in the case of the Iron-clad Vessels built by you, and now under seizure by H.M. Government, will be filed in a few days, and that it may be necessary to send a Commission abroad for the purpose of collecting evidence.

I am, Gentlemen,
Your obedient Servant,
(Signed) G. A. HAMILTON.

Messrs. LAIRD BROTHERS,
Birkenhead Iron Works,
Birkenhead.

APPENDIX.

CORRESPONDENCE

BETWEEN

OFFICERS OF HER MAJESTY'S CUSTOMS

AND

CAPTAIN INGLEFIELD, R.N.,

AND

MESSRS. LAIRD BROTHERS,

RESPECTING

THE IRON-CLAD VESSELS

BUILDING AT BIRKENHEAD.

Assistant-Collector of H.M. Customs, Liverpool, to Laird Brothers.

CUSTOM HOUSE, LIVERPOOL,
8th October, 1863.

Gentlemen,—Pursuant to the instructions I have received, I beg to transmit you the enclosed letter from the Lords Commissioners of H.M. Treasury, and to inform you that I have been directed to place a Customs officer on board the Iron-clad Vessel, now nearest completion, in the Great Float, Birkenhead, and that he has directions to seize her in case any attempt be made to remove her from where she is at present.

I am, Gentlemen,
Your obedient servant,
Signed) W. G. STEWART,
Assistant-Collector.

Messrs. LAIRD & Co.,
Birkenhead.

———————

Laird Brothers to Assistant-Collector of H.M. Customs, Liverpool.

BIRKENHEAD IRON WORKS, BIRKENHEAD,
8th October, 1863.

W. G. STEWART, ESQ.,
Assistant-Collector,
H.M. Customs, Liverpool.

Sir,—We beg to acknowledge receipt of your letter of this date, informing us that you have been directed to place an officer on board the Iron-clad Vessel now nearest completion, in the Great Float, Birkenhead, and that he has directions to seize her in case any attempt be made to remove her from where she is at present.

D 2

We have given Mr. Morgan, the Surveyor of Customs, an order of admission to the Iron-clad (which is named the " El Tousson") now lying in the Birkenhead Float, which order he will show to our watchman or ship-keeper when going on board.

We are respectfully,
Your obedient Servants,
(Signed)   LAIRD BROTHERS.

Laird Brothers to Assistant-Collector of H.M. Customs, Liverpool.
[*Private—further reply.*]

BIRKENHEAD IRON WORKS, BIRKENHEAD,
8th October, 1863.

W. G. STEWART, ESQ.

Dear Sir,—You have made a slight deviation in the wording of your letter of this date from that of the letter you sent over to us from the Treasury. You say, " has directions to seize her in case any attempt be made to remove her *from where she is at present.*"

The letter from the Treasury speaks of the "*dock or float* where she is at present.*"

Now, it is clear that the Harbour Master has power to move the berths of vessels in the dock as may best suit the working of the dock; and although we have requested Captain Hookey to give us as long a notice of his intentions to move the " El Tousson " as he can consistently with the working of the dock, yet we feel that this notice may be given at a time when we cannot inform you of it, as it may be out of office hours.

We therefore suggest that the instructions should only apply (as we understand the Treasury letter to be) in the event of an attempt being made to remove the vessel from the dock or float, and not to the mere moving of the ship under the orders and direction of the Harbour Master.

We think that Mr. Morgan understands this, but feel that in a matter of this importance it is right to let you understand clearly what we consider we have been called upon to do by your letter and the letter from the Treasury.

We are, Sir, Your obedient Servants,

(Signed)   LAIRD BROTHERS.

Assistant-Collector of H.M. Customs, Liverpool, to Laird Brothers.

CUSTOM HOUSE, LIVERPOOL,
8th October, 1863.

Gentlemen,—I beg to acknowledge the receipt of your letter of this day's date, stating that you had given to Mr. Morgan, Surveyor, an order of admission to the Iron-clad "El Tousson," and beg to thank you for the facility afforded by you to the Officers of Customs at all times.

I am, Gentlemen, your obedient Servant,

(Signed)   W. G. STEWART,
Assistant-Collector.

Messrs. LAIRD BROTHERS.

Assistant-Collector of H.M. Customs, Liverpool, to Laird Brothers.

[*Private.*]

CUSTOMS, LIVERPOOL,
9*th October*, 1863.

Gentlemen,—I have received your private note of yesterday, and regret that you should have the trouble of writing on the subject.

In speaking of the place where the Iron-clad is at present, I meant merely to speak of the dock or float where she is at present, and which I used as synonymous with these terms.

I am, Gentlemen, your obedient Servant,
(Signed)   W. G. STEWART,
Assistant-Collector.

Messrs. LAIRD BROTHERS.

———

E. Morgan, Surveyor, H.M. Customs, Liverpool, to Laird Brothers.

CUSTOMS, LIVERPOOL,
9*th October*, 1863.

Messrs. LAIRD.

Gentlemen,—I hereby beg to give you notice that I have this day seized the Iron-plated Cupola Vessel now lying in the dock attached to your premises, by order of the Commissioners of Customs.

Respectfully,
(Signed)   E. MORGAN,
Surveyor.

*E. Morgan, Surveyor, H.M. Customs, Liverpool, to
Laird Brothers.*

SURVEYOR'S OFFICE, CUSTOMS,
Messrs. LAIRD.   *9th October*, 1863.

Gentlemen,—I hereby beg to give you notice that, by directions of the Hon. Commissioners of Customs, I have this day seized the Iron-clad Vessel now lying in the great Float, Birkenhead.

Respectfully,

(Signed)   EDWARD MORGAN,

Surveyor.

---

*Laird Brothers to Captain Inglefield, R.N.,
H.M.S. " Majestic."*

BIRKENHEAD IRON WORKS, BIRKENHEAD,
12th October, 1863.

CAPT. INGLEFIELD, R.N.

Sir,—Understanding from you that you have received instructions from H.M. Government to take such precautions as you may deem necessary to prevent the Iron-clad " El Monnassir " (now being completed in our Graving Dock) from being forcibly taken away without our consent, and consequently nullifying the engagement which exists between us and H.M. Government in respect to this vessel, and as the vessel cannot be removed from our Graving Dock without lifting the caisson at the entrance, and thus affording free egress to the river, we hereby engage to give you reasonable notice of our intention to lift the caisson for the purposes of working our dock so that you may take such steps as you may think necessary to protect our property against the attempt which H.M. Government apprehend.

We are respectfully, Sir,

Your obedient Servants,

(Signed)   LAIRD BROTHERS.

*Capt. Inglefield, R.N., H.M.S. "Majestic," to
Laird Brothers.*

H.M.S. " Majestic," Rock Ferry,
14*th October*, 1863.

Gentlemen,—I beg to acknowledge the receipt
of your letter of yesterday, engaging to give me
reasonable notice of your intention to lift the caisson
of the Graving Dock in which the Iron-clad Vessel
" El Monnassir " is now being completed, and in
reference to our conversation yesterday regarding the
possibility of any of your workpeople being induced
to open the sluices without your cognizance, and by
which in one tide, the caisson might be floated out
of its present position, and the Iron Vessel thereby be
withdrawn into the river. I consider that your proposal that the keys whereby these sluices are worked
should be removed from the place they are at present
kept to another of greater security, under your personal care, is deserving of my thanks, and is again
suggestive of the good faith which has marked your
transactions with me in this unpleasant matter.
Allow me to take this opportunity of assuring you,
that as far as I have been informed, such has never
been doubted by those authorities who, for other
reasons have considered it necessary to place your
Iron-clad Vessels under the surveillance of the
Customs. I have only further to request that you
will let me be informed of your intention to open
your Graving Dock at least twenty-four hours before
the time proposed to float the caisson, and thus admit
of my making, by a personal interview, an arrangement for the security of your vessel.

Further, having a specific duty to perform, I beg you will not misunderstand me or imagine that I am actuated by a want of confidence in your assurances, should I find that at a later period it becomes my duty to absolve you from your present engagements to me, and take such other precautions as the then progress of the Iron-clad Vessel towards completion would justify. In the meantime I am satisfied that the present arrangements are sufficient, and (as you expressed to me) doubtless more convenient to yourselves than placing a party of men as a guard upon your premises.

I am, Gentlemen,
Your obedient Servant,
(Signed)　　E. A. INGLEFIELD,
Captain.

Laird Brothers to Captain Inglefield, R.N., H.M.S. "Majestic."

BIRKENHEAD WORKS, BIRKENHEAD,
14*th October,* 1863.

CAPTAIN INGLEFIELD, R.N., *H.M.S. "Majestic."*

Sir,—We beg to acknowledge the receipt of your letter of this date, in which you state—

That you have received our letter of the 12th instant, in which we engage, for the reasons enumerated therein, to give you reasonable notice of our intention to lift the caisson for the purpose of working our dock; so that you may take such steps as you may think necessary to protect our property against the forcible abduction which H.M. Government apprehends.

And further, that you understood from the conversation that we had yesterday regarding the possibility of our people being induced to open the sluices without our cognizance, and by which in one tide the caisson might be floated out of its present position, and the Iron Vessel thereby withdrawn into the river, that we undertook that the keys whereby the sluices are worked should be removed from the place in which they are at present kept to another of greater security, under our personal care.

We beg to inform you that we are quite prepared to confirm the engagement given in our letter of the 12th; but you are under a misapprehension in supposing that we undertook that the keys whereby the sluices are worked should be removed from the place where they are at present kept to another of greater security under our personal care. As we are not prepared to remove the keys of the sluices from under the care of the Superintendent of our Docks, in whose good faith and discretion we have implicit reliance; and we have given him special instructions to place the keys in a place of security under lock and key, which we know he has done.

With regard to the latter part of your letter, we offer no opinion as to the necessity or otherwise of the proceedings which Her Majesty's Government have taken, or may think fit to take, in relation to this vessel; nor do we admit that the engagement given by us is intended as an admission on our part that our arrangement for carrying out these proceedings is more convenient than another; but we undertake that we will give you reasonable notice of our lifting the caisson, through which alone egress can be had to the river, so that you may take such steps as you may think necessary to protect our property against the attempt which H.M. Government apprehends. And as you have informed us that you think at least twenty-four hours' notice is necessary to admit of your making, by a personal interview, an arrangement for the security of our vessel, we will endeavour to give you not less than this length of notice.

We are, Sir, your obedient Servants,
(Signed)   LAIRD BROTHERS.

Laird Brothers to Captain Inglefield, R.N., H.M.S. "Majestic."

BIRKENHEAD IRON WORKS, BIRKENHEAD,
19th Oct., 1863.

CAPTAIN INGLEFIELD, R.N., *H.M.S. "Majestic."*

Sir,—Referring to our letter of the 14th instant, we beg to inform you that we intend to open our Dock, in which " El Monnassir" now lies, on Thursday morning next, at about 7 o'clock, and also on Saturday morning at about 9 o'clock.

On Thursday the " El Monnassir" will not be moved out of Dock, but on Saturday she will be taken outside the gates to allow the Holyhead steamer " Alexandra" to pass out; after which she will be hauled into dock again, and the caisson will be immediately put into its place.

We are, Sir, your obedient Servants,
(Signed)   LAIRD BROTHERS.

NOTE.—We shall be glad to have an acknowledgment of the receipt of this as soon as convenient.
(Signed)   LAIRD BROTHERS.

Capt. Inglefield, R.N., H.M.S. "Majestic," to Laird Brothers.

H.M.S. "MAJESTIC," ROCK FERRY,
20*th October*, 1863.

Gentlemen,—I beg to acknowledge the receipt of your letter of the 19th instant, informing me that you intend opening the dock in which "El Monnassir" is now lying on Thursday morning next, about seven o'clock, and also on Saturday, about nine o'clock.

I am, Gentlemen, your obedient Servant,
(Signed) E. A. INGLEFIELD,
Captain.

Messrs. LAIRD, Birkenhead.

Laird Brothers to Capt. Inglefield, H.M.S. "Majestic."

BIRKENHEAD IRON WORKS, BIRKENHEAD,
22*nd October*, 1863.

CAPT. INGLEFIELD, R.N.,
H.M.S. "Majestic."

Sir,—With reference to our letter of the 19th instant, as we were unable to get a ship lying in our No. 1 Dock floated to-day, for the purpose of removing her to our No. 4 Dock—the one in which the "El Monnassir" is lying—we shall be under the necessity of opening this dock again to-morrow morning.

We are, Sir,
Your obedient Servants,
(Signed) LAIRD BROTHERS.

Please to acknowledge receipt of this per bearer. We regret having to give rather a shorter notice than you named to us, but we think it will be sufficient.

(Signed) LAIRD BROTHERS.

Laird Brothers to Capt. Inglefield, H.M.S. " Majestic."

BIRKENHEAD IRON WORKS, BIRKENHEAD,
24th October, 1863.

CAPT. INGLEFIELD, R.N.,
H.M.S. " Majestic."

Sir,—Owing to the dense fog, we were unable to open our dock this morning, but intend to do so on Monday morning.

Be so good as acknowledge receipt of this intimation.

We are, Sir,
Your obedient Servants,
(Signed)   LAIRD BROTHERS.

Capt. Inglefield, H.M.S. " Majestic," to Laird Brothers.

H.M.S. " MAJESTIC," ROCK FERRY,
26th October, 1863.

Gentlemen,—I have to acknowledge your letter of the 24th instant, acquainting me that you propose to open your graving dock to-morrow morning.

I am, Gentlemen,
Your obedient Servant,
(Signed)   E. A. INGLEFIELD,
Captain.

CUSTOM HOUSE, 28/10/63.
(*28th October.*)

Dear Sir, — We have received instructions to transmit to Mr. Bravay a notice of seizure similar to that which was handed to you yesterday.

Will you be so good as send per bearer the address of that gentleman.

Your obedient Servant,
(Signed)   E. MORGAN,
Surveyor.

Messrs. LAIRD BROTHERS.

BIRKENHEAD IRON WORKS, BIRKENHEAD,
28*th October*, 1863.

MR. MORGAN.

Dear Sir, — In reply to your note of this day, asking for Mr. Bravay's address, it is as follows—

Messrs. A. Bravay & Co.,
6, Rue de Londres,
Paris.

Your obedient Servant,
(Signed)   LAIRD BROTHERS.

Captain Inglefield, H.M.S. " Majestic," to Laird Brothers.

H.M.S. " MAJESTIC," ROCK FERRY,
28*th October*, 1863.

Gentlemen,—I beg to acquaint you that I have received from the Lords Commissioners of the Admiralty a letter, of which the following is an extract :—

" Desiring that full possession should be immedi-
" ately taken of the two Iron-clads now under seizure
" at Birkenhead, that Messrs. Laird's workmen should
" be immediately removed from them, and that the
" vessels themselves should be removed into the
" Mersey, and stationed where you may determine,
" with a sufficient guard placed on board of them."

I have therefore to request you will deliver the
vessels in question to my custody, upon my sending
an officer and party to take charge of them.

I am, Gentlemen,

Your obedient Servant,

(Signed) E. A. INGLEFIELD,

Captain.

Messrs. LAIRD, Birkenhead.

Laird Brothers to Capt. Inglefield, H.M.S. " Majestic."

BIRKENHEAD IRON WORKS, BIRKENHEAD,

28*th October*, 1863.

CAPTAIN INGLEFIELD, R.N.,
H.M.S. " Majestic."

Sir,—Referring to the conversation you had with
our Mr. John Laird, jun., this morning, and the
request you made to us for assistance in preparing
the " El Monnassir" for removal from our Graving
Dock, we shall feel obliged if you will put your
request in writing, and we will then give it our best
consideration.

We are, Sir,

Your obedient Servants,

(Signed) LAIRD BROTHERS.

Captain Inglefield, H.M.S. "Majestic," to Laird Brothers.

H.M.S. " MAJESTIC," ROCK FERRY,
28*th October*, 1863.

Gentlemen,—With reference to your letter of this date, requesting that I will ·put in writing the verbal application I made to you this afternoon for assistance in preparing the "El Monnassir" for removal from your Graving Dock, I beg now to renew the application, and request that you will give me such assistance by the loan of an anchor and cable, being essential to enable me to moor that vessel with safety in the Sloyne.

I have further to add, that such anchor and cable will be accepted as a personal loan. And I undertake that it shall not be considered as a part of the seizure of the aforesaid iron-clad vessel "El Monnassir."

I am, Gentlemen,
Your obedient Servant,
(Signed) E. A. INGLEFIELD, *Captain.*

Messrs. LAIRD, Birkenhead.

Laird Brothers to Captain Inglefield, H.M.S. "Majestic."

BIRKENHEAD IRON WORKS, BIRKENHEAD,
29*th October*, 1863.

CAPTAIN INGLEFIELD, R.N.,
H.M.S. "Majestic."

Sir,—We have the honour to acknowledge the receipt of your letter of the 28th instant, informing us that you have received from the Lords Commissioners of the Admiralty a letter, of which the following is an extract:—

" Desiring that full possession should be immediately taken of the two Iron-clads now under seizure at Birkenhead; that Messrs. Lairds' workmen should be immediately removed from them; and that the vessels themselves should be removed into the Mersey and stationed where you may determine, with a sufficient guard placed on board of them;" and that you therefore request that we will deliver the vessels in question to your custody upon your sending an officer and party to take charge of them.

We beg formally to protest against the illegal and unconstitutional seizure of these ships.

We shall, of course, offer no obstruction to the physical force with which we are threatened by the Government.

At the same time we protest against the probable destruction of our property in having ships, one of which is a mere hulk, without masts, funnel, or stearing gear, taken out of the Docks, where they are now in a place of safety, and moored in the river Mersey, at this inclement season of the year; and we trust that the Government will reconsider the orders they have given you on this point.

> We are, Sir,
>> Your obedient Servants,
>> (Signed)   Laird Brothers.

E

Laird Brothers to Capt. Inglefield, H.M.S. "Majestic."

BIRKENHEAD IRON WORKS, BIRKENHEAD,
29*th October*, 1863.

CAPTAIN INGLEFIELD, R.N., *H.M.S "Majestic."*

Sir,—We have the honour to acknowledge the receipt of your communication of the 28th instant, requesting us to render you assistance in preparing the " El Monnassir" for removal from our Graving Dock, and further to grant the loan of an anchor and cable, which are essential to enable you to moor the vessel with safety in the Sloyne.

We have every desire to render you personally any assistance in our power in carrying out the illegal and unpleasant duty imposed on you, but having given the matter very serious consideration, and regarding the responsibility we are under to the owners of those vessels, we greatly regret that we cannot, in justice either to them or to ourselves, do any thing to relieve H.M. Government from the responsibility they are under to us and to the owners in attempting to remove from our Graving Dock into the Mersey a vessel in the helpless condition of the " El Monnassir."

You are aware that in order to remove the " El Monnassir" it will be necessary to remove the caisson. This is an operation requiring some skill, and, in order to prevent injury to the caisson, we shall instruct our foreman to remove it on the day when you intend to remove our vessel.

We renew our protest to you at the illegal and extraordinary conduct of the Government in this matter.

We are, Sir, your obedient Servants,

(Signed)   LAIRD BROTHERS.

Capt. Inglefield to Laird Brothers.

H.M.S. " MAJESTIC," ROCK FERRY,
30th October, 1863.

Gentlemen,—At 10 o'clock to-morrow morning I shall send a Lieutenant, the chief engineer, boatswain and carpenter, to make a survey and inventory of the furniture and fittings of the " El Tousson." I hope this will suit your convenience.

I am, Gentlemen,
Your obedient Servant,
(Signed)   E. A. INGLEFIELD,
Captain.

Messrs. LAIRD, Birkenhead.

Capt. Inglefield, H.M.S. "Majestic" to Laird Brothers.

H.M.S. " MAJESTIC," ROCK FERRY,
30th October, 1863.

Gentlemen,—I have taken the advice of Mr. Bond, the Pilot, upon the subject of moving the " El Monnassir," and he states that it would be most imprudent to attempt to move that vessel into the river without a second anchor on board, unless she could be put to a buoy, and it will not be earlier than Monday that I can obtain the use of the latter.

I am obliged, therefore, to postpone taking that vessel out of your Graving Dock to-morrow, as I cannot get an anchor on board in time.

I can only add, that if this should cause you any inconvenience I am really much concerned.

And always,
Yours respectfully,   .
(Signed)   E. A. INGLEFIELD,
Captain.

Messrs. LAIRD, Birkenhead.

Laird Brothers to Capt. Inglefield, H.M.S. "Majestic."

BIRKENHEAD IRON WORKS, BIRKENHEAD,
31st October, 1863.

CAPT. INGLEFIELD, R.N.,
H.M.S. "Majestic."

Sir,—We beg to acknowledge receipt of your letter of the 30th instant, informing us that you have taken the advice of Mr. Bond, the Pilot, upon the subject of moving the " El Monnassir," and that he states it would be most imprudent to attempt to move that vessel into the river without a second anchor on board, unless she could be put to a buoy; that it will not be earlier than Monday that you can obtain the use of the latter; and that you are, therefore, obliged to postpone taking that vessel out of our Graving Dock to-morrow, as you cannot get an anchor on board in time.

In reply we beg to state that, owing to the lowness of the neap tides during the next week, we cannot with safety float the caisson at the entrance of the dock, and therefore must decline doing so.

We are, Sir,
Your obedient Servants,
(Signed) LAIRD BROTHERS.

Capt. Inglefield, H.M.S. " Majestic," to Laird Brothers.

H.M.S. " MAJESTIC," ROCK FERRY,
3rd November, 1863.

Gentlemen, — Obedient to instructions I have received from the Lords Commissioners of the Admiralty, I have removed the " El Tousson " from the Great Float, and moored her near my ship. It, however, became necessary to pay the dock dues that the ship might be cleared, and as you expressed to me when I spoke to you myself on the subject, that you did not intend to pay them, I have deposited the sum of £150 with the Dock Committee under protest, and now renew my request that you will pay these dues, so that I may be refunded the amount deposited.

I am, Gentlemen,
Your obedient Servant,
(Signed)   E. A. INGLEFIELD,
Captain.

Messrs. LAIRD, Birkenhead.

Vacher & Sons, Printers, 29, Parliament Street, Westminster.